ABOVE: Innocence abroad – a baby bat-eared fox;
David Rabinowitz.

THE BEST OF

AFRICAN WILDLIFE

STRUIK

Struik Publishers (Pty) Ltd
(a member of The Struik New Holland
Publishing Group (Pty) Ltd)
Cornelis Struik House
80 McKenzie Street
Cape Town 8001

Reg. No.: 54/00965/07

First published for CNA in 1988
Second edition 1993

10 9 8

Text © Struik Publishers (Pty) Ltd,
1988, 1993
Photographs © The photographers,
as individually acknowledged, 1988,
1993

Designer: Joanne Simpson
Dustjacket designer: Alix Gracie
Captions: Peter Joyce

Typesetting by Diatype Setting cc,
Cape Town
Reproduction by Unifoto (Pty) Ltd,
Cape Town
Printed and bound by Tien Wah Press
(Pte) Limited, Singapore

ISBN 1 86825 450 X

PHOTOGRAPHS:

FRONT COVER: Three cheetahs drinking
from a waterhole; *Lex Hes*.
TITLE PAGE: Dust-clouds and the din of
hooves – a wildebeest bull on flank duty;
Danni Suskin.
IMPRINT PAGE: At the going down of the
sun; *Jack Weinberg*.
BACK COVER: The king at rest – a male lion
in repose; *Lex Hes*.

The talent of the wildlife photographer, his skill, patience and artistic ability, captures on film the exquisite beauty, power and the grandeur of our wildlife heritage, allowing the talented and enthusiastic few to share their moving experiences with all mankind.

We need to reach out, not only to those fortunate ones who have experienced the magic of the wild at first hand, but to inform, stimulate and delight the eyes of a wider and less privileged audience.

It is up to this generation to preserve and protect the myriad and immense wonders of this earth, the seas and the skies. It is a holy legacy, entrusted to us for the benefit of our children, and those who follow after them. Through the miracle of photography, we can not only educate and delight, but, above all, heighten public awareness of the vital need to conserve our natural heritage for generations still to come.

PREVIOUS PAGES: The quiet dawn — a solitary giraffe stands sentinel among the dense, mist-wreathed vegetation of Timbavati, west of the Kruger National Park; *G. B. Keeping*.
OPPOSITE PAGE: Symphony in green and gold; *Reg Gush*. TOP LEFT: The glory of the paradise flycatcher; *Lanz von Hörsten*. TOP RIGHT: European bee-eater at breakfast; *Lanz von Hörsten*.
ABOVE: The lesser double-collared sunbird, common throughout southern Africa and often seen in city parks and suburban gardens; *Lanz von Hörsten*.

ABOVE LEFT: The paradise flycatcher, known to the Xhosa as Ujejane, has a short, sweet song, like the ripple of laughter; *Hein von Hörsten*. TOP RIGHT: Vlei, stream and marsh is home to the lovely malachite kingfisher; *J. Kloppers*. ABOVE RIGHT: The golden-banded forester butterfly; *A. J. J. Weaving*.

Swallow-tailed bee-eaters on colourful parade; *Koos Delport*.

Fantasia of frogs. OPPOSITE PAGE FAR LEFT: Sticking it out; *Trevor Wolf*. OPPOSITE PAGE ABOVE: Reed-frog trio; *Leen van der Slik*. ABOVE: So who's afraid of the dark? *P. J. Bishop*.

Family life. TOP: Two generations of chacma baboon; *David Rabinowitz*. ABOVE: Group portrait; *R. Levy*. OPPOSITE PAGE: Gregarious buffalo seek shade in the heat of the Zambesi Valley; *Keith Meadows*.

ABOVE LEFT: The shy waterbuck, which likes to seclude itself in dense riverine vegetation; *Paddy Hagelthorn*. TOP: Rusty spotted genet; *Fraser Smith*. ABOVE: The peaceable but, when wounded, cunningly dangerous buffalo; *Paddy Hagelthorn*.

Browsing bushbuck, with admiring audience; *Barrie Wilkins*.

The thirstlands OPPOSITE PAGE: Gemsbok, at home in dry country; *Louis B. Marais.* ABOVE: Skeletal dune-tree; *Trevor Skinner.* OVERLEAF: Swift retreat in the sandy wilderness; *Peter Franklin.*

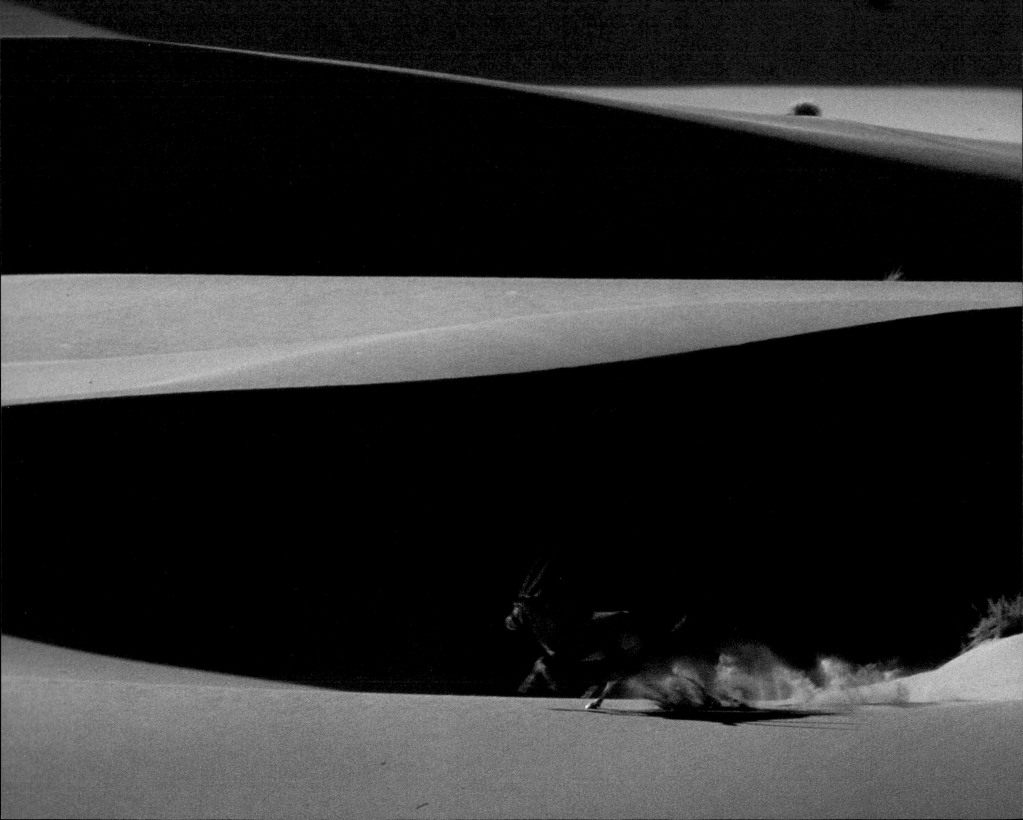

Images of the Namib. ABOVE: Sossusvlei; *David Rabinowitz*. OPPOSITE PAGE: Sun, shade, and an infinite silence – the magic of a secret place; *Gavin Thomson*.

OPPOSITE PAGE: **Twosome on trek;** *Trevor Wolf*. ABOVE: The young bachelors – springbok follow their lonely path to maturity in the immensity of the drylands; *Mike Reed*.

The desolate earth. ABOVE: Portrait of a survivor; *W. D. Haacke*. OPPOSITE PAGE: The long, long way to water; *Gavin Thomson*.

Raw Africa. ABOVE: Only the hardiest of plant life can withstand the Namib harshness; *Dudley Corbett*. RIGHT: The king of the desert drags away his kill; *Richard Goss*. OVERLEAF: Springbok of the great plains turn away from a storm; *Leon Nell*.

Lift-off: African spoonbills at sunset, Ndumu Game Reserve, Natal; *Reg Gush*.

The world of water. TOP LEFT: Study of a hamerkop, masterly builder of nests and an inspiration of fear in African lore; *Jack Weinberg*. ABOVE LEFT: A South African shelduck takes the plunge; *Trevor Wolf*. ABOVE RIGHT: In reflective mood – an open-billed stork; *J. Kloppers*.

Marine magic. ABOVE: Tsitsikamma gauntlet; *Patrick Wagner*. RIGHT: The acrobat of the ocean – a dolphin breasts the waves; *Patrick Wagner*.

ABOVE LEFT: The delicate filigree of a gorgon-head; *Ken Findlay*. ABOVE RIGHT: Battlestar ctenophora; *Ken Findlay*. RIGHT: The rock-hopper penguin, distinguished by its long, pale-yellow plumes, is a 'summer vagrant', found on the southern coast from Lamberts Bay to Pondoland; *John Visser*.

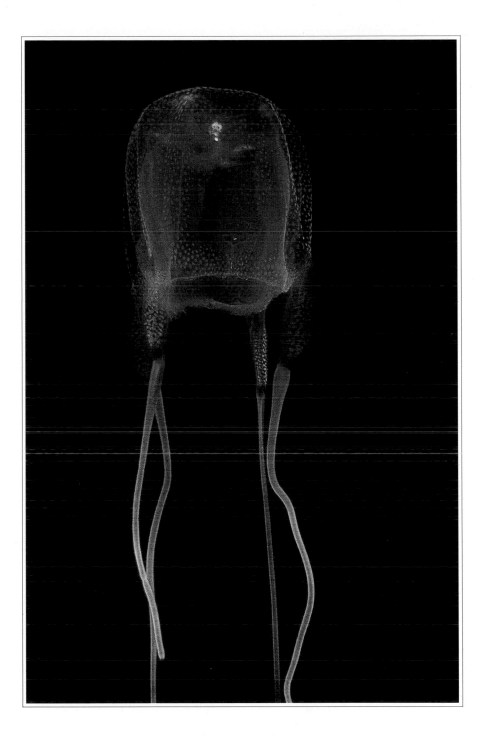

LEFT: Urchins of the sea woods; *Ken Findlay*. ABOVE: En pointe.
The venomous cubomedusa jellyfish; *Ken Findlay*.

Loving couples. LEFT: Hartlaub's gulls in the mating season; *J. Cloete-Kruger*.
ABOVE: The finely-marked Cape gannet, which gathers in huge colonies on the southern off-shore islands; *Lanz von Hörsten*.

The wonderful world of birds. ABOVE: A yellow-billed stork wades through the menu; *Wendy Freer*.
OPPOSITE PAGE ABOVE LEFT: The migratory white stork, large and with a distinctive red bill, arrives from northern climes in great flocks during November; *Roy Johanneson*.
OPPOSITE PAGE BELOW LEFT: The African darter, distinguished from the cormorants by its sharply pointed bill; *John Wesson*. FAR RIGHT: The sergeant-major – a grey heron; *Piet van't Riet*.
OVERLEAF: Lechwe in full and splendid flight; *Peter Johnson*.

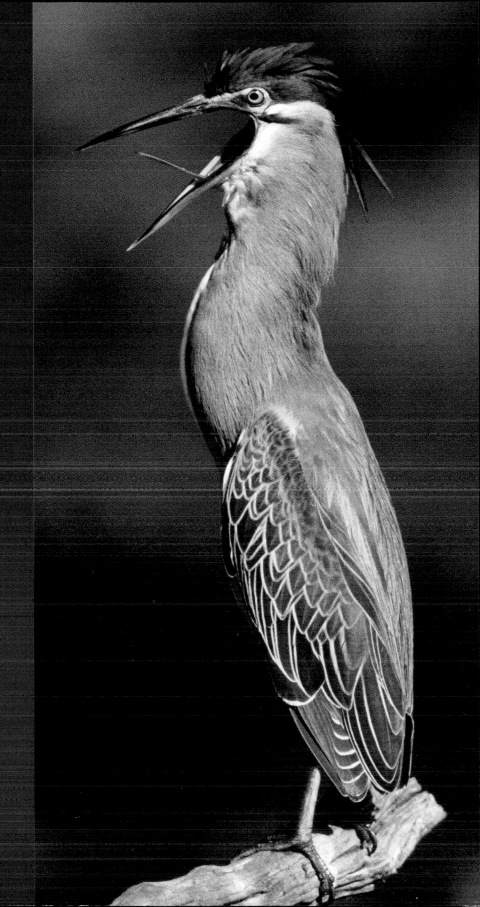

Zebra mêlée; *S. Jones*.

Prey and predator. ABOVE LEFT: A springbok slakes its thirst in the vastness of Etosha; *J. Kloppers*.
TOP RIGHT: The Nile crocodile, deceptively sleepy-looking; *Ronald Crous*. ABOVE RIGHT: Kudu mother
and its young; *Terry Carew*.

Proud mother and foal; *Koos Delport*.

50

A study in symmetry; *M. J. Davis.*

Sound and movement. ABOVE: A kudu watches, and listens, for signs of danger; *A. N. Jones*.
RIGHT: Burchell's zebra scramble for safety; *Mike Crewe-Brown*.

The competitive world of the male antelope. TOP: Springbok wrestle for dominance; *Louis Marais*. ABOVE: Gemsbok lock their splendid horns; *J. Kloppers*. RIGHT: Blue wildebeest in violent action; *Jill Sneesby*.

LEFT: Drama on the waterfront – an impala in macho mood; *J. Steyn*.
ABOVE: The end of an epic; *Michael Mills*.

Elegance and energy. TOP LEFT: A silver fox sights and jumps for a tasty morsel; *Michael Mills*.
TOP RIGHT: The 'pronking' springbok – the stiff-legged, high-leaping action is unique to the species, though its purpose remains unexplained; *Barrie Wilkins*.
ABOVE: Variations on the pronking pose; *David Rabinowitz*.

Like the springbok, the impala is a prodigious jumper, able to leap fully three metres high in a series of graceful bounds; *Norman Patterson*. OVERLEAF: Red hartebeest in flight; *Francois Maré*.

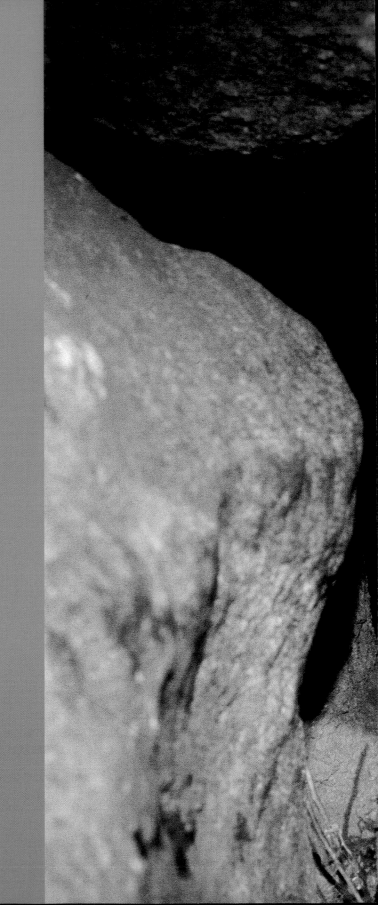

The two faces of the attractive, superbly agile caracal, a feline sometimes referred to as the desert lynx; *Lucius Moolman*.

The cheetah – swift, beautiful, and tragically endangered. LEFT: Pride personified; *Louis Marais*.
ABOVE: Nestling against mother's protective flank; *Brian Varaday*.

Sounding a note of warning; *Lydia Burger*.

Cheetah, able to achieve speeds of 95 km/h in short bursts, need open country for the chase and kill. These two enjoy their success; Photographers: *Ernie Buric* LEFT and *David Rabinowitz* RIGHT.

The adaptable leopard's diet consists of anything from field mice to large antelope.
OPPOSITE PAGE: The kill is hauled to the safety of high branches; *Lex Hes*. ABOVE: Siesta; *Ian Thomas*.

The much-maligned spotted hyena is not just a scavenger, but also a fine
hunter, able to confront and bring down prey as large as zebra and wildebeest.
LEFT: Hyena and hippo; *Trevor Wolf*. TOP: The moment of kill; *Michael Mills*.
ABOVE: Bloody mealtime; *Michael Mills*. OVERLEAF: Drama of the hunt; *Koos Delport*.

Silent, watchful and with an acute sense of hearing as well as sight, the owl is an efficient nocturnal hunter, its diet comprising small mammals, birds, frogs, snakes and insects. OPPOSITE PAGE FAR LEFT: The giant eagle-owl; *Greg Miek*. OPPOSITE PAGE ABOVE RIGHT: Taken by surprise; *A. Walter*. OPPOSITE PAGE BELOW RIGHT: Eagle-owl; *G. P. L. du Plessis*. ABOVE: White-faced owlets; *Barry Lovegrove*.

OPPOSITE PAGE: **The stark beauty of a desert evening;** *P. Smith*. TOP LEFT: **The double-banded courser, found in southern Africa's arid, sandy western regions, is largely ground-living, preferring to use its legs rather than its wings;** *Louis Marais*. ABOVE LEFT: **Kori bustard;** *Reg Gush*. ABOVE RIGHT: **Yellow-billed hornbill, common in the acacia and mopane areas of the north;** *Keith Ager*.

LEFT: Monarch of the mountains − a black eagle finishes building its lofty nest; *Peter Steyn*.
ABOVE: A pair of secretary birds settle down to roost; *Cecilia Boschoff*.

Two handsome raptors. OPPOSITE PAGE LEFT: A bateleur; *Greg Miek*, and ABOVE: The black-shouldered kite, the commonest and one of the most striking-looking hawks of southern Africa; *Len Miller*.
OPPOSITE PAGE RIGHT: Journey's end – the bleached remains of a buffalo; *J. C. Cunningham*.

ABOVE: Imperial presence; *A. Walter.* RIGHT: A juvenile bateleur eagle eyes its reptilian dinner; *Richard T. Watson*. Bateleurs are renowned for their aerial acrobatics.

A dog's life. The black-backed jackal is a resourceful scavenger, often competing successfully with hyenas for the remains of a kill. TOP: Confronting a tough-looking bateleur; *Ute Lucks*. ABOVE LEFT: Sharing the spoils; *Ute Lucks*. ABOVE RIGHT: A leisurely breakfast; *Mark Croft*.

Voracious vultures move in to feast, *David Rabinowitz*.

Sand, sea, and the ways of a scavenger. Black-backed jackals are common along the full length of the Namibian coast, hovering around fishermen's catch-landings or foraging wide for dead or dying marine life. OPPOSITE PAGE: On beach patrol; *Peter Tarr*. TOP LEFT: Precious water in a dry land; *J. Kloppers*. TOP RIGHT: Resting, but still watchful; *Louis Marais*. ABOVE: Picture of contentment; *Keith Ager*. OVERLEAF: Springbok in the great spaces washed with sun; *M. Bruce*.

Carefully protected survivors of an ancient breed. The white rhino is peaceable, despite its ferocious appearance. ABOVE: Fore and aft; *Leon Nell.*
RIGHT: Massive momentum; *Ray Hill.*

The gentle giants. The African elephant, prized for its ivory and needing more space than the modern world can afford, is declining in numbers to the point, in places, of regional extinction.
LEFT. Debut days; *Colin Richards*. ABOVE. On the run; *N. A. Smith*.
OVERLEAF: Watermarked; *Maurice Calvert-Evers*.

The harmony of nature. LEFT: Kings together — only man poses a threat; *Herman Staal*. ABOVE: A giraffe waits his turn at the water; *John Wesson*. OVERLEAF: Wildebeest in flight; *Peter Craig-Cooper*.

Lords of the veld. TOP: An afternoon stroll; *David Rabinowitz*. ABOVE: Mirror-image of strength and grace; *Roy Williams Jones*. RIGHT: The snarling prelude to love; *Jill Sneesby*.

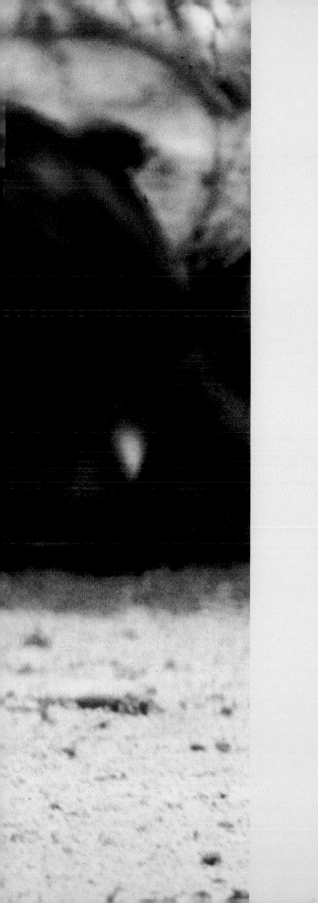

Lions in leisurely mood. LEFT: Curiosity is not likely to kill this particular cat; *Bev Joubert*. TOP: Out to the world; *Philip Perry*. ABOVE: A lioness begins its charge; *Barrie Wilkins*.

The killing time. RIGHT A co-operative effort; *H. H. Berry*.
ABOVE: Sharing the spoils; *H. H. Berry*.

Family life. LEFT: At the waterhole; *G. B. Keeping*. TOP: The young sentry *Trevor Wolf*. ABOVE: Two of a kind; *Jack Weinberg*.

Small world. TOP LEFT: The grey meerkat, or suricate – the very essence of innocent charm; *Marek Patzer*. TOP RIGHT: Family group of bat-eared foxes; *David Rabinowitz*. ABOVE LEFT: Squirrel secrets; *David Rabinowitz*. ABOVE RIGHT: Squirrel quarrels; *David Rabinowitz*.

TOP: Bat-eared fox on full alert; *Louis Marais*. ABOVE LEFT: Mongoose anticipating its meal; *Marge van't Riet*. ABOVE RIGHT: Overhead interest; *Henk Huizinga*. OVERLEAF: Springbok huddle together in a sudden, but welcome, rainstorm; *Madeleine Jacques*.

Hidden denizens of the Namib. ABOVE: A barking gecko breaks the desert silence; *W. D. Haacke.*
OPPOSITE PAGE: The perfect camouflage of a sidewinder snake; *Richard Watson*.

Focus on the reptiles. OPPOSITE PAGE TOP LEFT: The hidden eye; *W. D. Haacke.*
OPPOSITE PAGE TOP RIGHT: The green mamba, whose neurotoxic venom can be deadly;
Rob Nunnington. OPPOSITE PAGE BELOW: A young common tiger-snake emerges into the
world of light; *D. Heinrich.* ABOVE: Chameleon about to strike; *W. D. Haacke.*

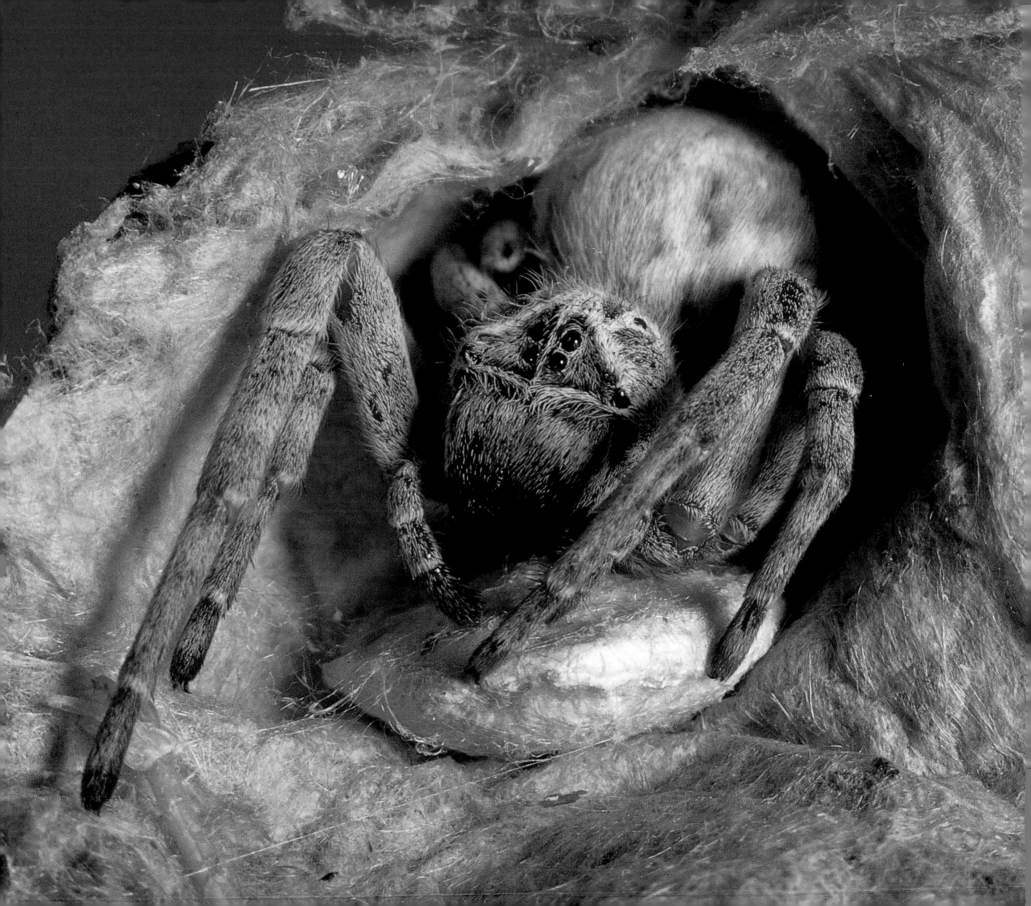

Arachnids and their eggs. OPPOSITE PAGE: Manoeuvring its egg sac; *Les Oates*.
ABOVE: Safeguarding the future; *Wendy Freer*.

The warthog: not one of nature's beauties, but tough, courageous and with a deep pride in family. ABOVE: With feathered friends; *Lex Hes*. OPPOSITE PAGE: Backing into the burrow; *R. du Toit*.

Golden evenings. ABOVE: Scavenging by last light; *Peter Tarr*.
OPPOSITE PAGE: Water-bed relaxation; *Ian Thomas*.

Descent of darkness. ABOVE: A chacma baboon on sentry duty; *David Rabinowitz*.
RIGHT: Night-kill; *Lex Hes*.

The end of an African day. OPPOSITE PAGE: A flamingo forages for its last snack; *Trevor Wolf*.
ABOVE: Waterbuck; *F. M. Hodgson*. OVERLEAF: Chobe evening; *Maurice Calvert-Evers*.

Homeward bound; *David Rabinowitz*.